1 8 9 6 ~ 1 9 9 6

Great Centennial Cattle Drive

Western Stock Growers' Association

Crew from last government horse round-up in the Bassano-Gleichen area, C. 1929. *Courtesy of Roy and Karen Clark.*

To the Western Stock Growers' Association's first 100 years
and
To our future ranching generations

1 8 9 6 ~ 1 9 9 6

Great Centennial Cattle 9W6 Drive

Western Stock Growers' Association

Produced by: John McQuarrie
192 Bruyere Street
Ottawa, Ontario
Canada K1N 5E1
(613) 241-1833

Distributed by: **Raincoast Books**
8680 Cambie Street
Vancouver, B.C.
V6P 6M9
(604) 323-7100

Design: John McQuarrie
Cover Design: Dave O'Malley
Camera-Ready Art: Bob Kingham
Colour Separations: Wellprint Inc, Toronto
Printing: Book Art Inc. , Toronto
E-6 Processing: Le Centre de la DIAPO

Canadian Cataloguing in Publication Data

McQuarrie, John 1946-
Great centennial cattle drive,
Western Stock Growers' Association

ISBN 0-9699761-3-5

1. Cattle drives–Alberta–History–20th century--Pictorial works. 2. Western Stock Growers' Association–Centennial celebrataions, etc.–Pictorial works. I. Title.

SF196.C2M33 1996 779'.99712303 C96-900542-3

Printed in Hong Kong
by Book Art Inc., Toronto

Working cattle north of Lethbridge. *NATIONAL ARCHIVES OF CANADA/PA-40574*

FOREWORD

In 1886 a cattle drive was anything but a celebration! It was hard work, long hours and a lot of worry – worry about the weather, getting enough good help, grass and water access and the market at the end of the trail.

Not much has changed in 1996 for Roy Clark and the team that organized the Western Stock Growers' Association Drive of 1996. So why would the WSGA choose a cattle drive to celebrate its 100th anniversary?

I believe it is because the stockgrower is proud of his heritage and the hard work required to make a living in an occupation that, for a century, has daunted lesser men and women.

Wild horse round-up near Sounding Creek, C. 1910.
NATIONAL ARCHIVES OF CANADA/PA-40604

Branding, C. 1905. *NATIONAL ARCHIVES OF CANADA/PA/40607*

The faces in these pages reflect the fierce pride of the independent cattlemen and women sharing the 100th anniversary of an association – and a *way of life* – with all those who wanted to be part of it.

The uniqueness of this cattle drive, and the pride of 100 years are vividly captured by John McQuarrie and Kim Taylor in the pages that follow.

Welcome to the Drive of '96 — an experience of a lifetime!

Dave Foat ,38th President
Western Stock Growers' Association

INTRODUCTION

Cattle sold. Horses returned to familiar summer pastures. Wagons safely back at the home place. Saddles and harness cleaned up and hanging in the tack room. And this great celebration of our first 100 years is now history. But what a celebration it was! It's times like this when we wish we hadn't tossed those superlatives around so freely. Because now that we really need them, they've been watered down some by over-use. Words like magnificent - wonderful - amazing come to mind. Just ask anyone lucky enough to have been a part of this Cattle Drive how it went, and those are the words they will use in describing their grand adventure. I've been back at the ranch for about a week now and it's still soaking in.

We've all heard that great old song; *Don't Fence Me In*. It's a cowboy's lament for the bygone days when he could ride from Calgary to El Paso without having to open a gate. By the time our Association was founded in 1896 fences had pretty much replaced night riders in holding cattle to one place. A hundred years later there was no open range left anywhere in the West. Well, almost anywhere. In 1941 the Government expropriated a large chunk of prime grazing country south of Buffalo, Alberta to facilitate a joint Canadian-British station where biological and chemical weapons of the day could be tested. After Hitler had been sorted out and chemical warfare outlawed, the British Army's tank regiments took over the place and did their thing.

Since fences and tanks don't mix, wire continued to be absent and this led us to the conclusion that this *British Block* would make one fine place to run 1500 people, 1700 horses, 2000 steers and over 100 wagons from Buffalo, 60 miles south to Medicine Hat in honour of our *Hundredth*. That's just what we did.

And I'm here to tell you that *it worked good!* Even though I, and quite a few other folks, are probably ten years older than we were five years ago. Most of the people who made this deal happen are listed in the back of the book. There isn't room here. I am one lucky son-of-a-gun to have had such a spirited, hard–working crew backing me up. I will never be able to thank them enough. And if you were on the trail with us – you will have the same trouble.

Someone once said that ranchers are the last to ask for help and the first to offer it. No better proof of this could be found than in the wonderful success of our:

Great Centennial Cattle Drive

Thank you one and all. Really!

Roy Clark
Cattle Drive Chairman

THE FIRST 100 YEARS

My neighbor Fred Burton on Meadow Creek always spoke of our Western Stock Growers' group as robbers. Like, "How'd things go at the Cow Thieves"? he'd ask over coffee. After any meeting in the past forty years I could, if I'd paid attention, answer with RCMP rustling reports, the latest on Bang's testing, freight rates, leases, taxes, tariffs, marketing boards and whatever the government had in store for us next. I used to think the "cow thieves" bit was in mock honor of the Stock Growers' early role as recorder of brands. Fred was a joker. But you never knew when he was being half serious either.

Fred was born to be a cowboy in 1898, but his parents were a homesteading couple and he would have been weaned on a folklore of old-time roundups careless with homesteaders' stock, "...scooping up any calf separated from its cow, accidentally or otherwise," to quote the Fort Macleod Gazette. Later, he might joke about greenhorn squatters and unbranded cattle, but he would also remember how the old roundup captains, "the greatest autocrats on the continent," in the words of one mounted policeman, often sold mavericks in the name of stock grower associations. Such was the custom when Fred's dad squatted on the Winder Ranche in the Porcupine Hills, one of the large "closed" leases of the day, and that custom was the subject of protest meetings at Fort Macleod and Pincher Creek until declared illegal in 1903 – after which an association director from Medicine Hat was jailed as a cow thief. So they say.

So maybe Fred wasn't joking. But it is relevant here that his young brother Gordon, also a squatter's son, was elected Western Stock Growers' president a generation later.

As Alex Johnston tells us in *Cowboy Politics*, our official 75th anniversary history, there were roundups in southwest Canada in 1879 and 1881 that would have required decisions on starting dates, holding areas, mavericks and so forth. But the first regularly organized roundup was held in 1882, following an order-in-council allocating much of Alberta and Assiniboia south of the Saskatchewan River to 100,000 acre "ranches" (spelled with an "e") to be leased for 21 years to responsible applicants. Such leases were closed to homesteaders, and the Fort Macleod Gazette plugged for a "strong compact Cattle Organization such as existed in Montana" to deal with inevitable problems. Thus, *The Pincher Creek Stock Association* was born. Ten years later the Gazette would be championing the homesteaders.

Ranche managers across the range began meeting to deal with wolves, Indians, rustlers, and sheepmen, and to organize the orderly recording of marks and brands. Thus evolved the *South Western Stock Association, the Canadian North West Territories Stock Association*, and both *Alberta* and *Southern Alberta Stock Growers' Associations*, all of which involved a very small handful of very large ranchers, most of whom had pull with Sir John A. Macdonald. As David H. Breen points out in *The Canadian Prairie West and the Ranching Frontier*, much of that influence was lost when the Laurier Liberals came to power in 1896, compelling stock growers from Maple Creek to High River to hang together. That led the smaller local groups to incorporate as the *Western Stock Growers' Association*, which represented cattle and horse owners across the southwestern Canadian range after 1896. Thus, the commemorative "9 Running-W 6."

As late as 1874, a doctor with the International Boundary survey stood on a promontory west of the Sweetgrass Hills and guesstimated half a million head as a buffalo herd grazed by. A couple of years later, and two hundred miles south, a U.S. cavalry officer thought he counted a million head in the Judith Basin. Our northern range was the last domain of the prairie bison or buffalo and just ten years later they were gone, surely not from slaughter alone, perhaps set back by diseases of the slowly encroaching longhorns. In any case, by 1880 the buffalo-dependent Indians, too, were on the edge of extinction. And here we ought to belay a conspiracy theory. If the White intention was to starve the Indians out, why trail in cattle to feed them? But that is what was done at enormous cost to the taxpayers ($500,000 a year, is a figure I've seen) to keep the Indians going while they learned a new way of life. Trail drives bound for distribution points from the Marias in Montana to Blackfoot Crossing on the Bow mark the beginning of our modern cattle industry.

Now imagine, as you browse this wonderful book, a buffalo range in breeding season. How would it have looked to hide hunters and bullwhackers? Like a "pristine" battlefield maybe? Scarred by cleat-like hooves, cratered by wallows, eroded near water, in places denuded and burned, with a thunderous bull-roaring over all. That's how I imagine our "British Block" military reserve in buffalo days. Nature, like War, is tough!

But so is Grass, and a season or two after the buffalo were gone, "needle-and-thread" and "blue grama" grass stems would have waved in the wind, giving rise to tales of a sea of grass waiting to be pruned by cattle. This book celebrates a way of life when cattle, sheep, horses and swine went to market on grass, and our 9W6 Trail Herd did just that, crossing a range reserved for war, unharvested and unpruned for fifty years, looking much as it did to those who saw it a few years after the bison were gone. There were problems on that range for the earliest cattlemen. In 1887, wolves were the biggest problem. Three years later it was "scabies" or mange, with which mite-infested livestock lost hair and died of exposure. This required a general roundup where mangy critters could be sorted off and dipped. Blackleg headed the list in 1901 and the Western Stock Growers' petitioned government to develop a vaccine at "reasonable" expense. It was also noted that illiterate railroad hands couldn't read brands. It followed that cattle killed on the CPR should be skinned and their hides officially inspected. In 1903, members worried that legal advice might be needed from time to time and they were of the opinion that a good attorney could earn his keep at $25 a day – for time actually spent on business.

With the 1914 War, wheat prices soared and there was no longer any legal way to exclude the plowmen. My neighbor on Cripple Creek, Ian McLeod – another homesteader's kid – remembered being posted with a washtub and a stick, drumming up a great racket and spooking wild range steers away from his family's oats. And, naturally, every homestead needed water, so water reserves were a very hot topic for rangemen. A 1914 meeting stressed the importance of farmers and ranchers working together, while at the same time considering the greater problem of "settlers squatting on... springs (formerly reserved) for cattle." In 1919 the weather and markets dried up and farmers left the country with the topsoil. While the last large, absentee-owned cattle companies folded with drought and depression, many tough homesteader families held on, producing crops of homegrown schoolteachers, cowboys and ranchers. By the 1950s, many a prominent "Cow Thief" bragged of squatter descent.

It is well to honor the trail drives and roundups of the open range. Imagine dozens of riders gathering thousands of cattle over hundreds of square miles and moving them to appointed places on time, without advantage of radio or cell-phone. That took planning! And our 9W6 Cattle Drive is proof that we still have the cowboy skills to do it today.

Another unchangeable is our time-honored freedom philosophy. The WSGA Newsletter for March 1996 quotes dead American president William H. Harrison ("Old Tippecanoe" he was called, and a Whig, as I happen to know because my ancestor was a member of his cabinet). He's supposed to have said "The only legitimate right to govern is an express grant of power from the governed," which sounds a lot like Pincher Creek rancher, F. E. M. Robinson, haranguing a 1948 Stock Grower's meeting on "...the right of any government to run our affairs."

John Schmidt, in his fascinating *An Experiment That Worked*, gives many examples of the thinking, as when president George Ross Sr. expressed the frustrations that followed an American tariff and domestic price-control boards. "Without the free right to market, western cattlemen have no future," said Ross, and he reportedly thought such restrictions might be reasonable cause for secession.

There's the yarn about Jimmie Gardiner, the Liberal Minister of Agriculture, who was "detained" by Stock Growers' members at Lethbridge until he promised to push his government hard for an open border. Was he plied with booze and hog-tied, or simply corralled in a room in the Marquis Hotel? (Such things would have been normal at a Stock Growers' "short course" twenty years later). But in 1948, manager Ken Coppock followed up on the open border issue with editorials in *Canadian Cattlemen* and reported "a flood of memberships" resulting.

But no one put the stock growers' philosophy better than Charlie McKinnon (president 1948-49), who wrote Mr. Minister Gardiner, "We are a free country of free men, and we favor free enterprise which has been responsible for the growth of this nation. We are opposed to state socialism... and will not desist in our battle for the right of private export." Charlie could speak for the industry. He had ridden with his father's LK wagon on the Bow River range, was building a successful ranch company and was on his way to becoming cattle feeder and meat packer. A respected WSGA member for sixty years, he became its first honorary Life Director.

250 years of experience. *Jack Hallet, Charlie McKinnon and Milt Ward 86th Annual Convention, Red Deer.*

But Charlie didn't invent the freedom philosophy. Chris Mills, who had been many places and seen many things before coming to WSGA, says that graziers are free enterprisers everywhere. He claims that the very nature of the business – always at the mercy of weather, requiring long-term commitments to large areas of land

Hereford cattle on southern Alberta ranch, *C. 1927.*
NATIONAL ARCHIVES OF CANADA/PA-40564

Trailing cow and her calf across Milk River. *Leslie Ranch, Picture Butte, C. 1944*
Harry Rowed /NATIONAL ARCHIVES OF CANADA/PA-133635

Round-up on a7 Ranch, *C. 1929.*
NATIONAL ARCHIVES OF CANADA/PA-40596

Trailing cattle through badlands, *C. 1944.*
Harry Rowed/NATIONAL ARCHIVES OF CANADA/PA-133636

attracts independent people opposed to the "strappings" of government. "I know of no country in the world," Chris told me, "where range livestock people are socialists. They can't afford the political interference and market instabilities that government creates. It's a chicken-and-egg kind of thing; a rancher's child sees what politics does to business at an early age. He or she doesn't need to be told: *Watch out for Government!*"

My purpose here is simply to introduce the outfit, not to mention every wonderful character who has contributed to its success. But we wouldn't have lived so long without a superb line of managers and secretaries: Eion Chisholm, Fred Newcomb, Chris Mills, Bernie Butterwick, Florence Causey, Debbie Brummond and Pam Miller, to name a few. In the years when we couldn't afford a secretary; the tradition was do-it-yourself and pass-the-hat. Thus in 1916, two years after absorbing the *Medicine Hat Stock Growers'*, funds and all, we were broke again, and volunteer manager Dr. Rutherford did just that – he passed the hat. The familiar issues were brands, wolves, the CPR, grazing leases, right-of-entry, water rights, and mange – especially mange.

Mange had been diagnosed in 1886, and it spread so fast that the entire western range south of the CPR was placed under quarantine. A coating of oil or a dip in a lime and sulfur concoction was known to control the mite. In 1899, a group of larger ranchers working through the Western Stock Growers' promoted a dipping vat at Kipp. They favored a dipping ordnance and forced compliance, which they considered reasonable restraints to freedom, since trans-border movements were threatened. But smaller ranchers resisted controls, saying their places were fenced, their cattle clean, and (I imagine) believing that the threat of mange was a plot. So, split on basic philosophy, our intrepid WSGA high-centred on mange, and the flow of dues dried up.

Cattle crossing Milk River en route to summer pasture, *C. 1944.*
Harry Rowed/NATIONAL ARCHIVES OF CANADA/PA-133639

Often it takes a crisis to get from lethargy to action and in 1919, crisis came with the mange. Then, one hundred and ten ranchers, under the leadership of George Lane of the Bar U Ranche and A. E. Cross of the a7 Ranche subscribed $5,000 each - an incredible sum - to start a new *Stock Growers' (sometimes Cattlemen's) Protective Association of Western Canada*, under the same old WSGA charter. They retained R. B. Bennett, a well known Calgary lawyer and later Prime Minister of Canada, to persuade the government to make dipping compulsory and to station the Northwest Mounted Police at each vat. That draconian action worked: mange seems to have been quickly eradicated; dipping vats were drained; quarantines were lifted; and after just four years the SGPAWC voted to change its name back to the *Western Stock Growers' Association*. John Schmidt reckons the episode brought lasting prestige, indicating a will to get things done - $500,000 is a lot of will! But fancy lawyers don't come cheap and when the smoke cleared away, WSGA (by whatever name) was again $7,000 in debt . Some things never change.

In 1923, Ruth Rogers-Peppard broke the gender-gap in the Stock Growers' office and stayed on as secretary/manager for thirteen years. Next in line was Jack Byers and in 1939, the controversial Ken Coppock took the reins. Journalist Frank Jacobs remembers Coppock as "a tremendous egotist but top-flight public relations man, whose talents the Western Stock Growers' should have used better."

As John Schmidt tells the story, Coppock was a well educated, ranch raised entrepreneur who had established a business at Cochrane. As Stock Growers' manager he understood that ideas have consequences and was never at a loss for ideas, but he needed a sounding-board and as usual, the directors had no money to invest in publishing. So Coppock put up $1000 of his own to kickstart *Canadian Cattlemen Magazine*.

Next, he started a supply company, and the directors voted him the profits of both enterprises in lieu of salary. The matter of actual ownership was "pretty informal" Jacobs says; he remembers financial statements where WSGA's subsidiaries were simply reported as "doing just fine." When Coppock and WSGA finally split, Coppock kept the supply company and he sold *Canadian Cattlemen* to *The Country Guide*. With Frank Jacobs as editor *Cattlemen*, as it came to be called, carried a monthly WSGA news column for twenty one years, a service now provided by Garth Cochran.

Coppock really distinguished himself by promoting a Basic Herd tax exemption in the post war years when herd increases were taxed as ordinary income. He also pressed for a western branch of the Council of Beef Producers, an Ottawa-based creation of the 1920s. So it must be said that Coppock served his employers well using the term loosely, since he usually went unpaid.

Next to "ride point" was Eion Chisholm, who has served WSGA as secretary/manager, director, president, and governor for over forty years. Who could better lead us to modern times? Applying for Coppock's job in 1953, Chisholm appeared at the unofficial Stock Growers' office in the Palliser Hotel. There he found vice-president John Cross stretched out on a bed in cowshit covered boots; president (and later Life Director) Bert Hargrave sat at a table with Charlie McKinnon, while director Eugene Burton from Medicine Hat propped himself up in a corner. Future president Clarence Copithorne sat on a stool in the middle of the room and gave him the third-degree; Eion could feel all their eyes burning right through him. "I was earning $3000 a year as a district agriculturist," he remembers, "and they were offering $4000 - a pile of money in '53 - and I took the job without even examining the books, which were dumped in the hall outside Coppock's office."

A director, Arthur Crawford-Frost, who was also a director of the Calgary Exhibition and Stampede, arranged a little office in the Stampede building, and Eion half jokingly remembers that, when he finally got around to inspecting the files, he found 90% of the members dead and the rest delinquent in dues for eight or ten years. The directors had talked of thirty five hundred members; come to find out, only three or four hundred were active. With his salary dependent on membership dues of half a cent per head of cattle owned (maybe $5 per member), Eion could see he would have to triple the membership just to earn his keep.

Meanwhile, needing revenue, he went for another store. There were very few vets in the country; drugstores were making big profits on blackleg vaccine, and other diseases were being diagnosed yearly, so Eion talked the directors into a second Western Stock Growers' Supply Company. His timing was right. Starting with $1000 working capital and next to no storage, the supply company financed office and staff, and many association projects, for twenty years.

Eion liked his new job but he needed something more for a decent living. So he and Thelma rented a little acreage west of Calgary – just a well, an unfinished house, and two corrals – and his president, Bert Hargrave, shipped him a load of steers.

"How do I pay for these steers?" Eion asked. "On the wages you're paying, I'm broke."

"You're not paying," said Bert; "just keep track of costs. If we make any money we split; if we lose any money we talk." That was the way it was in the 1960s.

It was also a milestone in Alberta's industrial history. For, while Chisholm served with little pay as WSGA manager, his directors, members and a few other cattlemen backed him while his feeding enterprise grew into Western Feedlots Ltd., the first real "custom" feedlot in Canada. Eion broke trail in an industry now active with commercial feedlots and world class packing plants, making Alberta the third or fourth largest cattle finishing region in the world today.

Meanwhile, the growth of the feeding industry pointed to a need for research, and one result was the CanFax marketing service. Born in the Stock Growers' office (and now a branch of Canadian Cattlemen's Association) CanFax ranks as WSGA's most productive offspring.

Foresight led the Stock Growers' board to promote and support the meat research of Dr. Howard Fredeen at Lacombe, which in turn led to major changes in a meat grading system that had long outlived its usefulness – one of WSGA's proud, if indirect, accomplishments.

But even with hardworking managers, change in the status quo doesn't come easily; often it requires lobbying. And in the 1950s and 1960s political lobbying was individual sport. As president, John Cross would take his problems directly to Dr. V. Wood, the wise and respected Minister of Lands and Forests, assuming that any solution that benefited him, would benefit the industry. And Wood, good politician that he was, would try to make sure that it did. Not a bad system really.

Interprovincial lobbying could be attempted through the Western Committee of the Council of Canadian Beef Producers, but contestants found that it usually paid to add colour. On top of his other duties, Eion Chisholm was appointed western secretary of the Council, and he describes some colourful lobbying as follows:

"After Diefenbaker was elected Prime Minister in 1958, it was decided to hold a Council meeting in Ottawa. We never expected to reach a minister or deputy; our highest hope was to get to some third-rate bureaucrat. That concerned us."

"What the hell's wrong?" Clarence Copithorne asked. "Why can't we get to the top?"

"Because we're nobodies," somebody said; "we just can't get their attention."

"So let's put on a western-style dinner," said Clarence; "invite our M.Ps., Senators, the whole Parliamentary Press Club. That should get some attention"!

"Let's do it!"—the board was unanimous.

Shipping Rocking P cattle at the Cayley stockyards. *Charlie Glas and George Chattaway.*

Don McKinnon was president of the Council, and he sent me down to Ottawa a week ahead to make the arrangements. Two cattlemen from each province were invited, but all the refreshments came from Alberta: thirty prime rumps from Burns & Company, beer from Calgary Brewing, vegetables from Lethbridge, whiskey from Alberta Distillers; we even took a western band down with us. They were sure gonna know we were there.

"How do we dress?" I asked.

"Like the Romans," Don said; "dress as they do; they'll all be in dark suits, white shirts and ties." But when we arrived at the ballroom, here's Prime Minister Dief in an old pair of Levis, big Stetson, yellow kerchief, and a pair of white goatskin chaps. Here's Minister of Trade and Commerce Hees in cowboy boots, and Alberta M.P. Jack Horner carrying a rope.

Well, the dinner and dance went ahead, and everybody got pissed like you wouldn't believe. George Hees grabs a mike: "I see one of you dudes with a rope. I bet nobody knows how to use it." Then, taking out his wallet: "Twenty bucks says nobody in the room can catch me."

Saskatchewan rancher, John Minor, took out an extra twenty. "Let's make it forty," he said. "Shall I catch ya by your head or your heels?"

"Oh, by the heels," says Hees, and he took off running down that big CPR hotel ballroom. Now John just flicked his rope and caught old Hees by both hind legs and flipped him, slick as a whistle. He didn't drop him hard, just upset him on that hardwood floor, but Hees went home with a goose egg and forty bucks lighter. Get their attention! That was the secret of lobbying in the '50s – and it's been downhill ever since.

Late in the 1950s it became the habit of a few Alberta ranchers named Ward, McKinnon, Harvie, and others, to attend Dr. Ensminger's Stockmens' School at Pullman, Washington, where they often found the University of Alberta's Roy Berg in the stable of lecturers. This not only made Berg internationally famous, but it sparked a homegrown effort, and in 1967 our own Cattlemens' Short Course was born. Sponsored by WSGA, planned by WSGA managers, administered by Professor Murray Hawkins and a U of A staff, taught by a world-class corps of entertainer/lecturers and appropriately housed at the famous Banff School of Fine Arts. Our short course was an instant success and became an annual event for fifteen years.

Eion Chisholm, Gordon Burton, Allie Streeter, Clarence Copithorne and George Ross Jr.

No one can know how many jokes, jugs, lessons and laughs were absorbed in those halls of learning. We do know that great and enduring ideas were hatched with the fun and fellowship. Practical lessons in animal breeding, nutrition, marketing and hedging went home with us. Ideas on "cowboy politics" whetted in all night bull-sessions have consequences in our capitols today.

One major consequence is Canadian Cattlemens' Association (CCA) raised from the bones of the Council of Canadian Beef Producers under bylaws drafted by Stock Growers' managers Newcombe and Mills in the WSGA office. Fred Newcombe went on at age 70, to become CCA's manager until relieved by Charlie Gracey. He then lived another thirty years in retirement, emerging only occasionally with advice on politics and women. Under Mills and Gracey, CCA flourished, raising the art of lobbying to a higher – if less colourful degree.

And when we saw what we had done in starting CCA, we were proud enough to want to give it teeth. CCA could lobby; given money it could also promote product, so the next step was a "check-off" to provide funds. But president Frank Gattey, visiting with Premier Harry Strom, learned that no check-off was possible without the approval of the Alberta Marketing Council. Several years and petitions and an ad hoc committee later, the Alberta Cattle Commission was born. Thus, CCA had money and WSGA had a rich relation, and hasn't been the same since.

In the late 1970s the Stock Growers', flushed with success, thought to channel its limitless energies (and very limited funds) into rancher-oriented banking, insurance, and real estate. What emerged under the leadership of president George Ward was the Stockmen's Center at 2116 27th Avenue NE, wisely built with outside capital, as a home for the family of Alberta livestock organizations. Manager Butterwick was installed in a prime location and other groups were lured in, including the infant Stockmens' Memorial Foundation and, for awhile, the Alberta Cattle Commission.

But ACC was soon regarded with suspicion. Authorized to collect a "check-off" (which seemed to grow each year, in the way of all taxes) and representing a very wide range of producers, perhaps the Commission couldn't be trusted to lobby for Stock Growers' interests.

A few weeks after taking office in 1978, I dropped in on vice-president Hugh Lynch-Staunton. The road in front of Antelope Butte is a smooth, if illegal, airstrip into which I occasionally dropped for coffee, and it may have been the very thick cream always found on Lynch-Stauntons' table (as we recently learned, fat is good for the brain)... anyway, something triggered the notion of an intelligent Stock Growers' "operative" (newspeak for "lobby-ist"), who could keep us abreast of goings-on in Edmonton. This was advanced in a newsletter and, with member support, the board agreed to an Action Fund (newspeak for passing-the-hat).

Stock Growers' director and M.L.A. Jack Butler nominated Sylvia Holcomb, whose knowledge of politics and people made her a shoo-in for the job. It was mutual love at first sight. "We are a hard-nosed bunch of sons-of-bitches," George Ward told her at interview.

"Then we'll get along fine," shot back Holcomb; "'cuz I'm a hard-nosed broad." And, with that, she became the Stock Growers' colourful intelligence operative, prowling the legislative galleries, feeding us facts, setting up committee meetings... Until, as was bound to eventually happen, the Action Fund went dry.

While we pride ourselves on George's hard-nosed description, the truth is that the modern WSGA is a family-based outfit that has championed family values for many decades, most of its members now being family ranchers. Modern conventions are more family-reunion than trade-show, and the place in our far-flung industry where new generations often meet the old. And we have "joined the nineties," as they say, with Pam Miller as manager; with Eileen

McElroy-Clayton, our first woman director, and Cherie Copithorne, trained in ranch management, serving on our board; with the team of George and Mabel Ward as Honorary Lifetime Members, and with the Clark family: Roy, Karen, and Rhonda; ramrodding our Centennial Cattle Drive.

Modern meetings of the "Cow Thieves'" are friendly, fun-filled, affairs. We remember Lethbridge in 1978, when the convention helped a horse-club for handicapped children with a bake-sale. Auctioneer Thelma Chisholm cried the sale and so confused the bidders that a plate of Amy Hargrave's cookies went for $80 a dozen – some kind of record – confounding bargain hunters like Tom Butterfield, Jim Lore, and Frank Gattey. When the cheering died away, it was determined that the hungry bidder was our honorary lifetime director, Amy's husband, Bert.

So, we celebrate our Centennial with a trail drive. A few old Stock Growers' hands still dimly recall when trailing was the only way to market. Trucks and trailers changed that, but beef cattle remain the same—self-propelled forage harvesters. That's a fact. And those of us who manage beef cattle commonly do so on horseback, herding from pasture to pasture in every kind of weather. That's tradition!

And the Honorable Jack Butler, still very much alive, first came to Alberta on horseback when the West was young. When the Bar U's fall work was done, he saddled up and rode back home to Manitoba – thankfully, to return for good the next spring. Well, most of us don't have as far to go, but most of us still use horses. We are Alberta ranchers – and that's tradition!

In our hundredth year, we still represent a peculiar breed of stock grower, the "rancher," who lives on a piece of land, a "ranch," and manages grass-covered "rangeland" for long term profit. And, believe it or not, a prime concern is keeping our environment healthy! That's good business! And tradition!

A rancher's freedom to manage property for production and for profit should become a public tradition if the prosperity of our country is to last. The hope and ambition of the poorest Canadian rests on the tradition of property. Think about it. Human nature!

Ranchers, with an eye to the weather, should pro-actively sell their traditional philosophy of individuality, freedom, and enterprise. Like the Cow Thieves' of old, a focused Western Stock Growers' should hold a roundup of ideas, "scooping up any ... accidentally or otherwise" that promote our way of life. Then we should sell those ideas *vigorously* to the public.

As our current president Dave Foat has written, "the Western Stock Growers' challenge is to share our vision of Individual Rights to Life, Liberty and Property with every citizen of Canada so that all may realize our traditional dreams and potential."

Sherm Ewing
Past President
WSGA

Sherm and Claire Ewing ranched in southern Alberta for 30 years and left 3 children and 7 grandchildren on the land. Sherm has written 2 books: "The Range" (1990); and "The Ranch" (1995), on the range environment and cattle industry history.

Time to move 'em out

Time to hitch the team.
Looks like a good day today. *Mert Taylor.*

Kim Taylor

Lead Captain. *Tom Osadczuk.*

JULY 20th

This staging area was grazing lease through the years, with such names acquiring it as the SS Ranch (Rheinhardt), the Anchor P (Macleay) and perhaps Gordon. Nearby were the Stapleton and Estes ranches. Estes became the Monarch ranch, owned by John Horne of Horne and Pitfield Wholesale Grocers. At various times horses, cattle and sheep were pastured here.

In 1941 Jim Wilson acquired the lease after his ranch was included in the "British Block'. After his death in 1947 it was owned by his sister, Mrs Helen McGuckin, who sold it to the Jones family in 1951. The lease is presently owned by Tom Osadczuk. In 1951, according to old lease invoices, lease rental was 9 1/4 cents per acre, with a stocking rate of 50 acres per head.

Immediately after crossing into the Base tomorrow, you will be in an area known as "Little Sweden" (so-called because it was thought that most of the settlers were Swedish, although actually they were of several nationalities.) The official name of the district was Ensleaf. Homesteaders began to come in 1912 and 1913, but by the time the area was evacuated in 1941 there were only a few famlies left. These were the Harvey Stone and Hazael Stone families, across whose land you will first travel; and the Rubbelke, Bartholow, Pedersen, Swenson, Hillsted, Barneko, Gillham, Hendry, Wilhelmson and Winkler families.

Heading out.

Following page: ***Wagons on the trail.***

Prickly Pear cactus.

Steers trailing south.

Kim Taylor

Herd's moving real good today. *Bud Maynard.*

Those that Have Gone Before

It seems fitting to pay tribute to those pioneers of the area of CFB Suffield who struggled against adversity to build communities and maintain their businesses. Not only the men should be remembered, for many wives and daughters took an active part in ranch and farm work. A debt of gratitude is also owed to scores of men who worked on the ranches through the years. They battled heat, dust, cold and blizzards, combined with what seems to today's generation extreme isolation and loneliness. They were "riders" for the early ranches, and when they rode on the roundups were called "reps" (representatives of the different ranches). Without them, ranchers could not have functioned. Some went on to other occupations, while others became successful farmers and ranchers in their own right

Expropriation, 1941

"In 1941, during WWII it was deemed necessary to establish a joint Canadian-British station where trials for biological and chemical defense could be conducted. The Suffield Block was chosen as the area had been declared unfit for agriculture under the Special Areas Act of 1938. By 1941 only 452 people remained from a high of 1649 in 1921."

"Dinosaurs to Defence, A Story of the Suffield Block" by BATUS and others.

Although the Act was signed in April it was early June before most of the residents received notice to vacate. The government compensated for buildings, and then gave permission to move them away. It was a major task to find building movers, or sell buildings if the resident was moving any distance, take down fences, make provision for livestock, etc. There are varying reports on compensation – ranging from $1.00 to $2.00 per acre for deeded land, 10 to 50 cents per acre for leased land, and payment on value of crops seeded. It was a traumatic experience for many who spent long years of hard work building up a home. It was impossible for most people to comply within such a short time, so it was later in the summer before the area was cleared.

Medicine Hat Stampede wagon heading home.

MEDICINE HAT
Est. 1887
STAMPEDE

Rainbow portrait. *Rae-Lynn Armstrong.*

Washing up. *Candace Massie.*

Shaving. *Ashmore Reesor.*

Roy and Karen Clark with Col Peter Kenward.

Following page: ***Drover's wagons heading for camp.***

Kim Taylor

Drover's conference.

JULY 21st

This camp, in the area of Dishpan Lake, was always ranching country. Large leases held by the W.D. McLennan family were known by such names as Brush Coulee Field, the Pocket Field and the Cold Storage, and covered nearly two townships. To the east was located the Arthur Orser family (1919), formerly the Ed Hubble site, and further east John Smith and family (1916) and, in early years, the Gallups. To the north-east, on what is shown on the map as White Horse Lake, was the John Wilson ranch, established in 1900 by James Black and bought by the Wilsons in 1905. The lake was always known locally as "Wilson's Lake". To the north of the Wilson ranch and just inside the "British Block" boundary the McGuckin family lived.

The trail from Medicine Hat to Battleford, Saskatchewan passed through here, going close to the Wilson ranch and continuing to a ferry crossing northwest of the present location of Bindloss.

About two miles east of here, and extending east and south to the South Saskatchewan River was Wawaskesy Park, an area set aside for the protection of the pronghorn antelope at a time when their numbers were rapidly diminishing. John Smith was for some time Game Guardian of the Park, which was always referred to as the "Antelope Reserve". The Park encompassed about 50 square miles.

Kim Taylor

Trail Boss. *Tom Butterfield.*

Discussing the route.

Drover returning to camp.

Kim Taylor

Kim Taylor

On to the next camp.

Camp.

Home on the range.

Another great day on the trail is done.

Rae-Lynn Armstrong

Peter MacArthur.

Laura Armstrong.

Greg Cruthers

Working on the Book. *Sherm Ewing.*

Pot of gold.

Rainbow over OK Corral camp.

Kim Taylor

Kim Taylor

Breakfast. *Lloyd Schlosser and Devin Cross.*

Refresher for the Wagon Boss. *Jim Bray, John Kalbhen, Ken Hoff and Leo Maynard.*

Wagon and harness detail.

Team's good-to-go.

Souvenier mug.

Coffee's on.

Night Herding

For two summers my cousin Norman, three years older than I, and I night herded the saddle horses. I would go on guard until 12:30, then call Norman and he would take over. When he saw the cook moving around in the morning, he would drive the herd in, place them in a rope corral and have breakfast. It didn't give two youngsters much time to sleep so, if we had a chance at noon, we could crawl under the wagon and sleep 'til called.

We were not allowed on the cattle herd at night at first because only experienced men were used there. After we had night herded the horses for two summers each was given a shift on cattle guard with a reliable older person. The flick of a match, the rustle of a slicker, a sudden noise such as a sneeze, or a sudden move could start a mild stampede. Singing or talking in a normal manner as you rode around them seemed to quiet them; they would lie down and chew their cuds but be always on the alert for a sudden change of any kind. Those on guard had to know when to let the cattle spread a little to feed and when to bunch them again to bed down.

Night herding was not a bad job so long as the weather was fairly good; we took it as a matter of course. The outfit was camped one night eight miles southwest of the ranch when one of those blustering thunderstorms came up. The thunder roared, lightning flashed, almost continuously; the cold rain, driven by a very strong wind, was like a flail and of course the darkness was black except when the lighting laid everything bare. One of those on guard with the cattle came to camp and called for all hands to get out and they were out till morning, mainly because when the storm did stop they weren't sure where they and the cattle had drifted to. (Each rider had a saddled horse tied to the wagons for use when the guards were changed.) Only the cook and I were left. The tent blew down and the beds were soaked but the grub was safe in the grub box.

"True Tales of Days That Are Gone (Rhymes of the Range and Other Poems)" by John R. Gordon

Kim Taylor

Camp visitors.

Alberta sunrise.

Bed roll wagon.

Checking the tool kit. Neil McKinnon and Vern Lonsberry (trial run '95)

Fresh horses.

Head-to-tail tie. *Joe Chase.*

Supply run.

Freighting Supplies

John R. Gordon recounted tales of trips across the area this group is traveling. Not for them a caterer and mobile communications. When making trips from their ranch on the Red Deer River to Medicine Hat for supplies, their camping equipment consisted of a tent and a bed roll (a tarpaulin fashioned into a sleeping bag with snaps and rings and lined with blankets.) A frying pan and a ten pound lard or syrup pail for making tea served as their only cooking utensils. The grub box contained simple fare – bacon or canned corn beef, bread, canned corn or tomatoes and perhaps a can of fruit.

The trip was usually made with four horses on an open, double-boxed wagon and took two days going in and three coming out with loaded wagons. With a good team on a democrat it was a day and a half trip, although some fast horses did it in a day.

Wagons under the big sky.

The way it was.

Picking up the drag.

Following page: ***Alberta blue.***

Wagon detail.

Wheel hub.

Kim Taylor

Lantern.

Kim Taylor

Stove pipe.

Rolling up the rope corral.

Kim Taylor

Drover's wagons rolling into Drowning Ford camp.

A little roping practice. *Brandy and John Kalbhen, John Naismith and Wes Clark.*

Watering the horses. *Rhonda Clark (trial run '95).*

How the Drowning Ford Got Its Name

Around 1855 a band of Crees were camped near the present Drowning Ford Ranch, while a band of Blackfeet were on the north side, out towards Brutus, hunting buffalo. The Crees, all young bloods, made a raft, crossed the river, and stole about thirty head of ponies from the Blackfeet camp, running them back to the river and swimming them across behind the raft.

On discovering the theft the Blackfeet followed the trail of the "horse–thieves" down to the river bank. The Crees having no better sense, flashed a message over with the aid of a mirror, never dreaming that the Blackfeet could cross so early in May, just after the breaking up of the ice. They started a war dance to rub in their victory. The Blackfeet, apparently, gave up the crossing of the river as hopeless and turned back on the trail they had followed to the river. However, when out of sight, about seventy of them went upstream about ten miles and swam their ponies across. About dark, they sneaked down and closed in around the Cree camp. At dawn they raised their war whoop and charged in among the Crees who, taken completely by surprise and outnumbered more than two to one, lost about a dozen Indians shot down by arrows as they came out of their tents.

Of the remainder, a number were drowned trying to cross the river, and only four of them escaped. The scene of this encounter and tragedy is known in the Cree tongue as the Drowning Ford.

"Early History of Medicine Hat" by J.W. Morrow as printed in "Wagon Trails to Pioneer Tales" Bowell to Bowmanton Historical Society, 1993

Kim Taylor

Watering the stock.

Watering horses.

JULY 22nd

Near this campsite was the Thomas Tinney ranch, located here in 1898 and continuing until 1941. Further along on the north bank was Billy Little who had a stone house.

The Lokier family lived at a location several miles south-west from about 1900 to 1912 when they moved across the river from here to establish Riverbend Ranch. Thomas Lokier developed a lovely ranch site, well planted with trees and even fruit trees! As early as 1903 he was experimenting with growing alfalfa as winter feed for cattle – at that time it was not usual to provide supplemental feed to cattle on native pasture. In 1917 a gas well was drilled which supplied power for pumping water to irrigate the alfalfa and ranch grounds.

South of here, William Winterburn had ranching holdings along the river, and to the west Jack Sewell ranched. Several miles to the west, in a settlement called Bingville, homesteaders began arriving about 1910 and continued for several years. By the 1920's many had left because of adverse farming conditions. By 1941 there were probably 20 to 25 families living there.

Kim Taylor

Afternoon water hole.

Kim Taylor

Kings Royal Hussars and the cowboys.

Kim Taylor

Kim Taylor

Kim Taylor

Kim Taylor

Brits and the Cowboys.

Wagon pulling into Hogs Back camp.

The Boys. *Morgan Chattaway, Allan Lively and Scott Chattaway.*

Bit.

Canteen.

Kim Taylor

Pickin' and grinnin'.

Kim Taylor

Swappin' yarns.

Harness detail.

Ready to hitch. *Clay Chattaway.*

Kim Taylor

Down the road.

Wagon Boss. *Leo Maynard.*

Over the hill.

Heading for a rest spot.

Following page: **Looks just like 1896 out here today.**

Cattle Drives and Roundups

We always had a few hundred head of our own cattle on the ranch but the large herd which was there for nine years and numbered, at one time, approximately 12,000, belonged to Gordon, Ironside & Fares Co. Ltd. The 'Firm' had no interest in the ranch or the horses. These belonged entirely to us. We were paid a certain sum per head, per year for running their cattle...

Toward the last of June each year we shipped a train load (at least twenty cars) of five and six year old steers (18 to a car) to England. These steers received no supplementary feed, either in summer or winter, except as calves. There was an abundance of cured prairie grasses, excellent shelter in the coulees and along the river, and no fences to prevent the cattle drifting with a storm until they found shelter. Another lot was shipped the last part of September. This shipment went to Winnipeg where it was re-sorted and part re-shipped. It consisted of five and six year old steers that had been missed in June, or were not in condition for export at that time (to be fit for export a steer had to have "his britches tied"), a few top four-year-olds to fill out a car, and some dry cows and bulls which had been on the ranch long enough.

Cattle were wild in those days and, once located on a range, needed no one to tell them when a storm was coming; they themselves were the best of weather prophets. If you saw the cattle working toward shelter you could be sure that a storm was brewing and when you saw them leaving the shelter and climbing the hills to graze, even if it was cold, you knew that better weather was on its way.

It always took at least three days to move the beef herd to the stock-yards at Stair (in later years at Suffield) starting from a camp a few miles south of the ranch. There would be 350 to 400 steers in the herd. Shorthorn, Hereford and Shorthorn-Hereford cross – almost everyone horned. When they were being gathered on the range there

Single file.

Nature's gifts.

would be some great fights as the different small bunches came together, meeting perhaps for the first time in a year.

The trek to the railroad through wide open country, making a new camp each night, night herding both cattle and saddle horses, was just part of the day's work for men who were in the saddle day after day, from May until October. The nights were usually cold in September. Often when you awoke in the morning, if you slept outside of the tent, your hair would be frozen to the tarp. When you were dressed you took your towel, went down to the slough or spring, broke the ice near the edge and washed the sleep out of your eyes. Then, after rolling your bed, you were ready for hotcakes, slices of bacon (not these thin shavings that you get today), biscuits and coffee. If you wished any, there was always a pot of cooked prunes, cooked dried apples, cooked dried peaches or perhaps a mixture of dried apples and apricots.

The train of empty stock cars was "made-up" in the Hat and, on shipping day, moved up to Stair or Suffield with the Brand Inspector on board. We always tried to have the cattle in the stockyards before the train arrived. If there was a misunderstanding regarding the time of arrival, the crew and any spectators were asked to stay in the caboose out of sight until the cattle were in the pens. As I said, they were wild and a small thing could make it almost impossible to corral them. When the cattle were loaded the train moved back to town where another crew, with their engine and caboose, took over. In those days stock trains moved at the same rate of speed as a passenger train. When they arrived in Winnipeg the cattle were unloaded, fed, watered, reloaded and shipped to Montreal where they were placed on a cattle boat bound for Liverpool, England.

The shipper was required to send at least one man with each shipment, both on the train and on the boat. The man was given free passage by the railroad to destination and his return ticket was purchased at a rate below regular fare.

Lined out and on the move.

The last of the herd.

The foregoing was excerpted from John R. Gordon's "True Tales of Days That Are Gone (With Rhymes of the Range and Other Poems)". In 1896, the same year that the Western Stock Growers' Association had its beginnings, Andrew Gordon (John's father) established the first ranch on the Red Deer River in this vicinity, a few miles north and west of the Buffalo staging area. Andrew Gordon's brother, J.T. Gordon, held a major interest in the firm of Gordon, Ironside & Fares Co. Ltd., which operated a meat-packing plant in Winnipeg and had various ranching holdings. A note of interest – at one time the Firm brought 2500 head of Longhorns from their New Mexico ranch, but they were not equal to the rigors of Alberta winters on the range and most did not survive.

Greg Curry.

Kim Taylor

Bill Atkins.

Kim Taylor

Jeff Taylor.

Kim Taylor

Joe Warren.

Kim Taylor

Valentina Gabert.

Bud Maynard.

Mary-Jane Block.

Paul Picotte.

David Pope.

John Kalbhen.

Kim Taylor

Threatening skies.

Drover's camp, Hogs Back.

Starting a new day.

Move 'em out!

Reminiscing.

JULY 23rd

In early years Purdy ranched in the area of tonight's camp and there were homesteads to the west. A few miles south of here and across the river was the Drowning Ford Ranch established by John Spencer. The barn was built with lumber that had been aboard the "Lily", a steamboat that ran aground near the Drowning Ford Ranch site. The Grant brothers and Dan Newton followed and in 1927 Harry Minor bought it. He built a 14 room house on the site. They also had a gas well and irrigation but all was taken over in the 1941 expropriation. An area east of the river which was originally included in the "British Block" was some years later returned to civilian use and is now part of the Drowning Ford Grazing Association.

Further east of the Grazing Association is the Bull Springs Ranch owned by the Hargrave family. Tom acquired it and later his son Bert managed the operation. Bert is a Past-President and honorary Life Director of the WSGA and he served as a Member of Parliament. The ranch remains in the family.

Kim Taylor

Good day coming. *Don Lowe.*

Guarding the herd.

Prairie renegades.

Greener pastures below.

Wranglers bringing drover's horses into camp.

Horse Roundups

The following is an account of horse roundups as remembered by Bert Shannon, whose first encounter with one was in 1912 when he was a boy, and was drafted to drive the bed wagon! The Shannon ranch was on the Red Deer River near Empress.

"It seemed always to fall to the Shannons to run the roundup wagon. Uncle Les (Shannon) was the wagon boss. It took about three weeks to make the circle. The roundup travelled up the Red Deer River to about Jenner and from there south–west to Tilley–Suffield. The riders would cover about ten miles on each side of the wagons which had the main herd. When I met the roundup at Suffield they had about 20 cowboys and from 800–1000 head of horses. This was quite a sight moving across the prairie, which in those days was quite open.

They were camped at Suffield for a couple of days cutting out the stray farmers' horses they had picked up in transit that had mixed with the range stock. From there the roundup moved east through what is now the British Block to the Wilson Ranch. The Social Plains area was usually gathered from the Wilson Ranch. Here the various brands were separated and each ranch took their horses and went their separate ways."

In the rope corral. *Roger Beierbach.*

Wrangling horses.

Kim Taylor

Home stretch.

Old and the new.

Crossing bridge into Medicine Hat.

No turning back.

Medicine Hat's finest.

Kim Taylor

Up the hill and no wrecks.

Kim Taylor

Full house.

Looking them over. Roy Clark & Craig Wiechnik.

Two more steers for the chutes.

Kim Taylor

Cutting steer for sale.

In Memory
Jeff Kalbhen

October 19, 1970 – July 25, 1996